Kelp Forests

JUDITH CONNOR AND CHARLES BAXTER

MONTEREY BAY AQUARIUM®

Monterey, California

The purpose of the Monterey Bay Aquarium is to stimulate interest, increase knowledge and promote stewardship of Monterey Bay and the world's ocean environment through innovative exhibits, public education and scientific research.

Acknowledgements We thank our friends and colleagues whose discussions, criticisms and inspiration helped shape this book: Steven K. Webster, Mike Vassar, Dick Zimmerman, Nora Deans, Chris Harrold, Jim Watanabe, Eugene Haderlie, Lovell and Libby Langstroth and the folks at Sea Studios.

Published in the United States by the Monterey Bay Aquarium Foundation, 886 Cannery Row, Monterey, CA 93940-1085.

Library of Congress Cataloging in Publication Data
Connor, Judith, 1948-
Kelp forests.

(Monterey Bay Aquarium natural history series)
1. Kelp bed ecology. 2. Kelps. I. Baxter, Charles, 1927-   . II. Title. III. Series.
QH541.5.K4C66 1989   574.5'2636   89–14553

ISBN 1-878244-01-9

Photo and Illustration Credits:

Balthis, Frank S.: 56 (bottom)

Bucich, Richard: 54

Caudle, Ann: 6 (art), 14, 39 (right), 40, 41 (bottom), 44, 46 (bottom), 50 (art)

Connor, Judith: 21

Davis, Chuck: 25, 42 (top), 50 (top left), 62 (top left)

Foott, Jeff: 8, 19, 36 (top right), 55 (middle), 61 (top left)

Gohier, Francois: 51 (top left)

Gotshall, Daniel W.: 31 (top left)

Hall, Howard: 11 (top), 17, 18, 27 (bottom), 32 (right), 35, 47, 59

Herrmann, Richard: 20

Hobson, Edmund: 57

Horn, Helmut: 10 (top)

Johnson, Geoffry: 4, 60

Kells, Valerie: 12

Langstroth, Libby and Lovell: 27 (top), 29 (left), 33 (bottom two), 34, 36 (top left), 37 (bottom left), 39 (top left), 42 (middle), 45, 48 (bottom), 53 (top), 58, 62 (bottom right)

Lanting, Frans/Minden Pictures: 31 (bottom right)

Monterey Bay Aquarium: 28 (bottom), 32 (left), 55 (bottom), 56 (top)

Nicklin, Flip and Assoc.: 26, 50 (bottom)

Rigsby, Michael: 52

Sea Studios, Inc.: 13

Snyderman, Marty: 37 (right), 48 (top), 51 (right)

Tucker, Karen: map-16

Watanabe, Jim: 61 (bottom left)

Webster, Steven K.: 10 (bottom), 13 (bottom), 15, 28 (top), 30, 31 (top right, bottom left), 33 (second from top), 38, 41 (top), 43 (bottom), 46, 55 (top), 61 (top right), 62 (top right) 63

Wrobel, David: 7, 9, 24, 43 (top)

Wu, Norbert: 11 (bottom), 22-23, 33 (top), 37 (top left), 29 (right), 42 (bottom), 43 (middle), 49, 50 (top right), 53 (bottom two),

Front cover photograph: Marty Snyderman
Back cover photograph: Mike Johnson
Back cover illustration: Ann Caudle

Series and Book Editor: Nora L. Deans
Designer: James Stockton, James Stockton and Associates
Printed in Hong Kong through Global Interprint, Santa Rosa, CA, USA

CONTENTS

A̲t the Monterey Bay Aquarium, visitors discover a new view of the ocean at the Kelp Forest exhibit. This exhibit is the only one of its kind in the world, a real centerpiece of the aquarium. As you enter the aquarium, you can nearly feel and taste the ocean nearby, but it's hard to prepare yourself for that first sight of a living kelp forest. Drawn by glimpses of blue and green, you approach the tall cathedral-like windows. Standing before a wall of windows two stories tall, you experience the living, swaying kelp forest.

From your diver's-eye view of the exhibit, the seaweeds, invertebrates and fishes inside show the real texture of marine life. Seaweed spores drift in with the sea water pumped from the bay. Settling down on the rocks, they start to grow. The developing plants compete for sunlight. Taller plants overgrow and shade those below. Towering golden fronds of giant kelp spread a canopy of blades across the surface of the water. Shafts of sunlight penetrate thin spots in the canopy to highlight the scene below where sea stars and anemones muscle in for their share of space on the rocks. Fishes rest on the bottom, swim through the seaweeds or hang motionless in the jungle of tall kelp.

Come back year after year, and you'll see a different scene—richer and richer, just like the kelp forests in the bay. It's a teeming, ever-changing forest community, more complex than you'd suspect at first glance. Each level of the kelp forest is home to a different set of animals, and, in nature, every kelp forest has its unique combination of inhabitants.

The breathtaking beauty of kelp forests along the Monterey coast inspired this book. From the towering golden kelp plants that structure the community to the whales and worms that pass through, the kelp forests are ripe for exploration.

1

KELPS

Scramble along the rocky shore at low tide and crouch down in a respectful bow to slippery rocks and seaweeds. Reach out and touch the tangle of nearshore kelps: tough cabbage heads, slick leathery neckties and feathery boas. Out beyond the pounding surf, you can see the canopies of the elegant deep water kelps. Isolated floats of bull kelp bob in the surge like the heads of swimming otters. Stretching upward from deeper rocks, a snarl of giant kelp spreads fronds across the water's surface.

Sea lettuce grows quickly in calm bays.

ALGAE, SEAWEEDS AND KELPS The giant kelp swaying in the kelp forest is just one member of the group of plants called algae. Algae are simpler than most land plants, which have roots, stems, leaves, flowers and other adaptations to life on land. Algae have no leaves, no stems, no roots and no elaborate reproductive structures. The term "algae" embraces plants in salt water and fresh, in soils, sand and even snow. Algae range in size from tiny, microscopic cells to the enormous kelps that grow over 100 feet long. In the oceans, some algae live permanently adrift in the water, while others settle down on rocks, sand or mud or secure themselves to marine animals or other plants.

Red orbs of sea grapes bob below the surface.

Seaweeds are algae that are large enough to see without a microscope and flowering grasses that grow in the oceans. The seaweeds include unrelated groups of plants: the green algae (Chlorophyceae), the red algae (Rhodophyceae), the brown algae (Phaeophyceae) and flowering plants, like surf grass (*Phyllospadix* spp.).

Green algae are similar to flowering plants in color and chemistry. Some, like bright green sea lettuce (*Ulva* spp.), really are weeds; they quickly overgrow rocks to crowd out other plants.

The brown Cystoseira *is related to rockweeds.*

Red algae are beautiful and diverse. There are more kinds of red algae—about 4,000 species—than any other kind of seaweed, but it's difficult to identify them by color alone. Some, like sea grapes (*Botryocladia pseudodichotoma*) and the calcified corallines (Corallinaceae), maintain a red or pinkish hue, but many intertidal red algae, like iridescent *Iridaea*, look purple or green.

Most brown algae have a golden brown or greenish color. They range in size from tiny, threadlike feathers to the thick-fingered rockweeds (Fucales) and the large, leathery kelps (Laminariales). The kelps are actually one small group of large brown seaweeds that make a big impact on the environment with their remarkable size and complexity.

A snarl of feather boa kelp and oarweeds envelops intertidal rocks along the California coast (right).

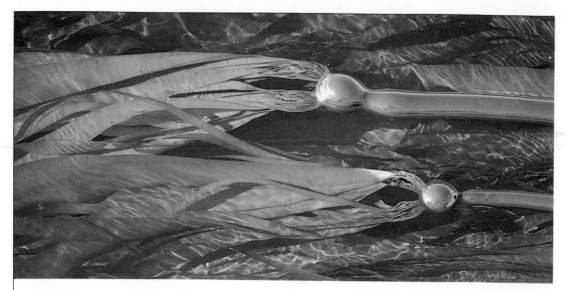

At least 20 different species of kelp grow along the coast of California. Some kelps are subtidal giants; others are shallow water streamers. All together they're a group of magnificent brown algae, a group that contains the largest and fastest-growing of the seaweeds. And because all kelps are similar in form, function and life history, scientists group them together in the order Laminariales.

Three basic parts make up the body of a kelp plant: the holdfast, stipe and blade. From hanging on for dear life to securing sustenance and sharing it, each component does its part to perpetuate the living kelp.

Attached to buoyant floats that lift them to the surface, golden blades of bull kelp stream out in the sunlit waters (above).

A rootlike holdfast anchors the giant kelp to a boulder on the seafloor (right).

THE KELP HOLDFAST An anchor against the surging waves, the holdfast secures the kelp in place on a rock. Rootlike but not a root, it can withstand the pull of currents and surging waves. The holdfast doesn't gather water and nutrients from the soil the way roots do; it just holds tight to the rocks. The holdfast develops from stubby branches that sprout from the base of the plant. These sprouts grow down over the rock surface and wedge into any available crack or crevice.

As the holdfast grows, branches continue to sprout from the base. The new branches intertwine and overgrow the older ones, giving the holdfast a conical shape. Small crabs and other invertebrates move into the spaces between the branches. In time, some members of this community eat the older, inner branches and hollow out the cone. Snails, brittle stars and worms join in, taking refuge inside cavities between the branches. In its prime, the spacious holdfast provides a secure home for a multitude of creatures.

With age, the holdfast loses its grip. Hungry tenants may overeat, or the oldest branches in the center may decay and fail. If the newer holdfast branches don't hold fast, surging storm waves will pull the weakened holdfast free, setting the kelp adrift. The kelp's secure life ends when its anchor slips.

THE KELP STIPE

The kelp's stipe is like a stem. Tough, but supple, it can whip in the waves without breaking. It's both a sturdy anchor line and a vital conduit between the holdfast below and the blades above. Inside the stipe, long trumpet-shaped cells aligned end to end compose a conducting system similar to that of land plants. It's through this conduit that sugar alcohols—the product of photosynthesis—travel from the upper blades in sunlit waters down to the dimly lit lower portions of the plant. The sugar alcohols move from the source, where there's excess, to less-productive parts, where there's not enough. This conducting system gives kelps their unique advantage among the algae: by sharing sweet assets from top to bottom, kelps grow larger than other kinds of seaweeds and overshadow their competitors.

Some kelp stipes develop enlarged hollow floats called pneumatocysts (pronounced "new-**mat**-o-sists"). A mixture of gases fills each balloonlike float, providing buoyancy. The float pulls the kelp blade toward the surface of the water where it can harvest more sunlight for photosynthesis.

Young floats buoy these columns of giant kelp, while older floats slump in decline (top).

A top snail rasps food from the stipe as it crawls along (above).

THE KELP BLADE The flattened blade is the real workplace of the kelp. In the blade, cells absorb water, carbon dioxide and other chemicals from their ocean surroundings and, using energy from sunlight, convert these simple chemicals into oxygen and food compounds—sugar alcohols, amino acids and other building blocks of the plant. These compounds provide nourishment for all parts of the plant, from the growing tips in the canopy to the established holdfast and developing fronds far below the surface. The stipe helps transport food to where it's needed most. Surplus food is stored in the cells of the kelp plant, making blades, stipes and holdfasts nutritious meals for countless kelp forest fishes and invertebrates.

Photosynthesis, this art of using sunlight to make food, sustains the plants that support almost all the living creatures on our planet. Sometimes the link from plants to herbivores to carnivores is direct, like the food chain from the kelp to a grazing snail to a snail-eating fish. When worms, crabs and other invertebrates, fishes, birds and marine mammals enter the picture, the entangled food chain is more aptly described as a food web.

Just as important as their role in providing food for animals today, some of the blades produce spores, the promise of future generations of kelps. By making spores, the blades initiate a new phase in the secret life of the kelps. The life history of the kelps is an incredible tale, as powerful as fiction.

THE LIFE HISTORY OF GIANT KELP The seaweed we recognize as giant kelp (*Macrocystis pyrifera*) tells only half the story; it's one of the two distinctly different plants that make up the life of this kelp. The larger plant, called the sporophyte, produces spores, while a different plant, the sexy little gametophyte, actually makes eggs or sperm.

Few people realize that the large kelp plants you see in the bay—the giant kelp, the bull kelp, the feather boas—are the spore-producing or sporophyte plants. Some kinds of kelp sporophytes, like the bull kelp, produce spores on any blade, but the giant kelp develops clusters of special fertile blades called sporophylls near its holdfast. Dark patches on these fertile blades disclose the location of many spore-making cells called sporangia. Each sporangium, smaller than the eye can see, produces 16 or 32 spores. Spores are the kelp's miniature units of reproduction. A kelp plant in its prime produces hundreds of sporophylls that generate billions of spores.

Spores are tiny, each smaller than the dot of a period on this page. On the kelp blade, they're tightly packed in the sporangia. Released when they're ripe, spores show their true form; they're lively kidney-bean-shaped swimmers. A swarm of the swimming spores travels from the kelp plant. Moving with the waves and currents, the swarm may travel a few feet or several hundred yards away. Most of the spores are lost at sea or eaten by animals; a few lucky ones find themselves stranded in calmer water near the bottom and settle down.

In sunlight, the blades of giant kelp become dynamic food factories converting simple chemicals in the sea into food and oxygen.

Inflated like a balloon, this swollen bull kelp float holds a curious mixture of gases, including carbon monoxide.

Once in place on a rock, each spore once again looks like a simple cell. Slowly over the next few days it divides to form a tiny chain of cells. This small chain is the other phase of the kelp's life story: the gametophyte phase with separate male and female plants that make sperm or eggs.

Compared to the huge kelp sporophytes, the delicate little gametophytes are much smaller, but they too have a subtle sex appeal. These plants produce the kelp sex cells, or gametes; half are male plants and half are female. Stimulated by blue light that filters down through the water, the tiny female plants produce eggs and the male plants make microscopic sperm. Chemical signals emanate from an egg in the clasp of the female plant. Released from the male plant, sperm swarm toward the female's egg, irresistibly drawn by its chemical perfume. For the sperm, it's a race to reach the female first. Only one sperm can successfully unite with the egg. The fertilized egg becomes a new sporophyte and a new generation of kelp begins.

Still attached to the female, the fertilized egg begins to swell, then divides in half. Fused together, the two halves select their separate growing assignments. The lower nub takes a downward branching course; the cells multiply and grow to form a delicate holdfast. The opposite end grows upward and flat like a simple knifeblade. Between the holdfast and the blade, a cylindrical stipe takes shape and grows longer to give the young plant greater stature. Within weeks, the sporophyte youngster overgrows its gametophyte mother and assumes a paddle shape. To this stage, most kelps follow the same storyline, but now each species takes its separate way.

Giant kelp, for example, stretches aloft to earn its name. A split develops where the blade and stipe meet. Starting as a small central rip, the wide blade tears lengthwise into parallel strips. Before that first separation is complete, new splits open that will render the two blades into four, then eight. Tiny gas-filled bulbs develop at the junction of each blade and stipe. The bulbs mature into buoyant floats, the pneumatocysts, that hoist the succession of blades into an upright frond. The stipes get longer, sending the fronds upward like a progression of pennants fluttering up a flagpole. Each frond of giant kelp flaunts the history of its development: older, full-sized blades below and smaller, younger ones aloft. A scimitar blade crowns each growing frond, cleaving itself into new miniature blades in sequence.

Six to twelve thin and spindly fronds grow upward from one giant kelp holdfast. These early fronds are self-reliant, each growing as fast as its own photosynthesis allows. Other parts of the plant won't be so independent. The holdfast and the new fronds developing in the shadows depend on sugar supplements from the fronds above.

While there's cooperation within the plant, there's stiff competition outside. Like rival siblings, the holdfasts of young sporophytes crowd the rock. The fastest-growing plants shade out their comrades to win the race for sunlight and space.

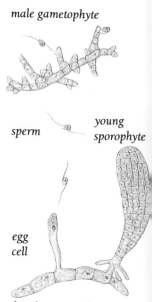

male gametophyte

sperm

young sporophyte

egg cell

female gametophyte

Tiny cells that crown the male kelp plant release their microscopic sperm. Alert for perfume from nearby eggs, sperm swarm down toward a female plant that bears a double burden. One long egg cell stretches up, while another egg, already fertilized, develops into a giant kelp on the mother plant.

After a year, a 100-foot-tall mature giant kelp presides over the domain it may occupy for seven years or more. This giant works prodigiously, producing food to keep young fronds growing and older parts going. Besides adding to its own visible assets, the kelp shares its wealth by leaking nutritious fluids, sloughing off blades and shedding fronds. The early fronds are already gone, their six-month lifespans over. New fronds, some with thicker stipes, have replaced the original fronds. The plant has produced a handsome canopy of blades and a dense cluster of sporophylls flourishes near its holdfast. These fertile blades release billions of spores to replay the kelp's cycle of life again and again.

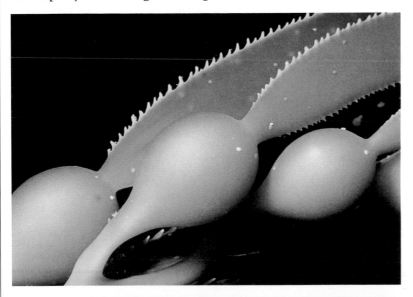

As the young giant kelp blade develops, small spikes protrude along its margin (left).

A scimitar blade crowns each frond of giant kelp forming tiny, new blades in a series from the tip (below).

KELP KIN

Can you tell the kelps apart just by their shape? In all, there are four families of kelp in California: Chordaceae, Laminariaceae, Lessoniaceae and Alariaceae.

Chorda, a shoestring of a kelp, is the sole member of the family Chordaceae and isn't found in Monterey Bay. But a variety of members from the other three kelp families abound in the bay.

You can tell the kelps in the Laminariaceae family by their plain and simple fronds with branched or cuplike holdfasts. Their blades fulfill all functions, making sugars and spores. In some species, the original blade tears lengthwise into strips, while in others it remains whole.

Kelps in the family Lessoniaceae usually branch out into complicated forms. These kelps have branched stipes, multiple blades and sometimes floats. Some of the kelps in this family develop special spore-making blades, while others grow spores right on their regular blades.

Members of the kelp family Alariaceae look feathery, growing secondary blades along the stipe or on the edges of a central blade. These feathery kelp grow their spores on special blades.

Giant kelp is a relative of bull kelp.

Dictyoneuropsis *is in the family Lessoniaceae.*

Bull kelp belongs to the family Lessoniaceae.

Feather boa kelp is in the family Alariaceae.

Alariaceae includes the feathery Pterogophora.

Laminariaceae claims the classic oarweed.

THE KELP FAMILIES   Each of the kelp families has its own characteristic form, but the environment plays a role in its shape and texture, too. Some kinds of kelp stay single, living as individual oarlike fronds; others divide to form countless, delicate tongues. Every kelp plant, regardless of species, starts out as a simple blade, but a small area at the base of the blade dictates the ultimate shape of the plant. In this area, actively dividing cells segregate into different tissues; they may branch to produce many blades or inflate to form bladders. The branching patterns of young plants are obvious; wear and tear, age and injury disguise the patterns in older plants.

Each species of kelp goes through life with a shape and style of its own. The plain and simple kelps in the family Laminariaceae persevere on wave-swept rocks. The oarweed (*Laminaria farlowii*) keeps its basic paddle shape, while *Hedophyllum sessile* broadens into leathery cabbage heads. In rough waters, the ribbed *Agarum* develops into a ruffled blade with lacy holes. The broadest oarweed (*Laminaria setchellii*) rips into a waving handlike shape, its upper edge frayed and tattered by the surf.

Kin to the bull kelp and giant kelp found in deeper waters, this sea palm thrives on inter-tidal rocks swept by the roughest waves.

Other kelps, including those in the family Lessoniaceae, elaborate on the simple, classic *Laminaria* design with dissected blades. Some, like the bull kelp (*Nereocystis luetkeana*), have gas-filled floats that buoy them on the surface. From its narrow holdfast end, the bull kelp's stipe extends 30 to 60 feet, gradually expanding to form a single large round float at the surface. Blades sprout from the top of the float and hang down like balloon streamers. Another member of the family, the giant kelp, produces a whole series of blades, each with its own little float. A venerable, seven-year-old giant kelp may stretch out 150 feet long, producing thousands of blades and floats in its lifetime.

Many members of the Lessoniaceae family are smaller, inter-tidal seaweeds. A look-alike of the giant kelp, *Macrocystis integrifolia* dresses deep, quiet tide pools with its many blades and floats. The rubbery sea palm (*Postelsia palmaeformis*) doesn't need buoyant floats. In an upright slouch, its thick stipe supports a crown of blades. Clustered high on wave-washed rocks, sea palms hunch together to face the surf.

The winged kelps in the family Alariaceae start life as simple *Laminaria* shapes, later developing smaller secondary blades along the sides of the main blade. Some of the smaller blades are sporophylls, specialized to make the spores for future plants. Near shore, a long fringe of the feather boa kelp (*Egregia menziesii*) twists and uncoils in the surge. Like a huge feather, *Alaria marginata* flaps its sporophylls in the waves. Growing even taller and more abundant in the rich, cold waters off Alaska, plumes of *Alaria* ascend from depths of 20 feet to form a floating canopy on the ocean's surface.

2

THE KELP FOREST HABITAT

Holding fast to the rocky bottom, giant kelp stretch upward, then spread their blades across the water's surface, creating a dense forest canopy. These golden canopies decorate the coasts of several continents, extending along the west coast of North America from Alaska south to Baja California, and along the temperate shores of South America, South Africa and southern Australia.

Like a barrage of rockets, a speeding school of jackmackerel bursts into the peaceful kelp forest (right).

What makes a particular patch of ocean suitable for growing kelp? The giant kelp's needs are subtle, but definite. It must have a hard surface for attachment, plenty of nutrients, moderate water motion and cool, clear ocean water. Giant kelp grows best in well-mixed salty waters near the outer coast. Poor water circulation and freshwater runoff inhibit kelp, discouraging its growth in stagnant bays and estuaries.

DISTRIBUTION OF MACROCYSTIS

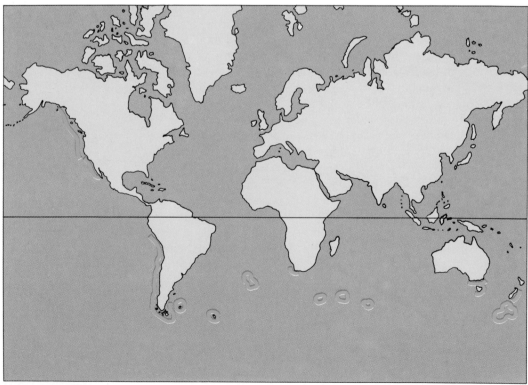

Macrocystis plants grace the coldwater coasts of North and South America, South Africa, Tasmania, Australia, New Zealand and the islands near Antarctica.

ATTACHMENT A rock in shallow water gives the giant kelp a place to hang its holdfast. The vertical contours and security of a kelp's attachment site depend on the character of local rocks, their resistance to erosion and the energetic forces of waves and swells. Along the coast of California, a wide variety of rocks offers some form of security.

Hard, durable rocks like granite give kelp a solid foundation and provide spectacular vertical profiles. Softer rocks, like siltstones, sandstones and shale, present low-relief contours and less secure attachment sites. The weakly bonded layers of these softer sedimentary rocks separate easily. Honeycombed by burrowing clams and mussels, they often fracture under stress. In the heavy surge of winter storms, there's always the danger that chunks of rock covered with kelp plants and other attached organisms may be torn loose.

Sand and silt cover most of the seafloor surrounding the more stable rocky outcrops where kelp forests develop. These loose sediments make precarious attachment sites for kelp; there's no stability in shifting sands. Sand has a destructive split personality, alternately abrading and then smothering holdfasts and their inhabitants.

Occasionally, sandy habitats provide a temporary respite for kelps. In sheltered waters, kelp plants that settle on small shells or worm tubes can extend their holdfasts out into sand and grow. As new plants settle on holdfasts of the old, a kelp forest develops. The sand and shale outcrops along the Santa Barbara coastline once supported dense kelp forests, protected from southerly ocean swells by offshore islands. Many of these kelp beds have disappeared, uprooted by unusually large storms that accompanied the "El Niño" weather pattern in 1983; no one knows how long it will take for them to recover. Some kelp beds may never recuperate if shifting sands keep kelp spores from settling down.

NUTRIENTS Like all plants, kelps use light energy to convert simple chemicals into food for survival and growth. Most seaweeds absorb the simple chemicals they need directly from their seawater surroundings; water, carbon dioxide, nitrate and phosphate provide the major necessary elements of hydrogen, oxygen, carbon, nitrogen and phosphorus. Traces of other essential elements, like manganese, iron, zinc and copper, are also absorbed from sea water. It's a dynamic exchange: seaweeds and phytoplankton drain nutrients out of the immediate neighborhood; waves and currents help mix them back in.

Normally, dissolved nutrients are abundant along the coast of central California, continually renewed by the upwelling of deeper nutrient-rich water and other ocean activity. In southern California, calm or unusually warm periods stifle ocean movement locally and nutrients become limited almost every summer. During those periods, giant kelp plants slow down or stop actively growing as the plants rely on traces of nutrients in the sea water to sustain life.

Rarely found in the colder waters of Monterey Bay, an orange garibaldi and a brilliant pink gorgonian decorate this southern California kelp forest.

During unusual "El Niño" weather, warm ocean currents carry exotic characters, like this pelagic red crab, from the warmer waters of Mexico north to Monterey Bay.

COLD WATER Although it's too cold for human comfort, kelp forests grow best in oceans where temperatures remain below 70°F. The large sporophyte can withstand temperatures up to 74°F, but the tiny male and female plants perish when water temperatures rise above 68°F.

Does the water have to be so cold? The answer isn't clear. Nutrients seem to disappear when ocean water warms up and that's when the kelp forests decline. Perhaps at high temperatures, kelps can't get the simple chemicals they need for photosynthesis and they starve. Maybe the warm temperatures simply signal that upwelling hasn't occurred to replenish nutrients in the water.

In southern California, kelp forests follow a regular cycle of development, growing best during cold periods in winter and spring, then deteriorating when waters warm up in summer and fall. Extended periods of warm water conditions, like "El Niño" weather, may cause long-term damage to the kelp forests.

CLEAR WATER Even when it's bright and sunny on the surface, there isn't always enough light for plants under water. For all the sunlight pouring down on the ocean's surface, only a fraction reaches the seafloor. Inside the kelp forest, the canopy itself shades out 90 percent of the sunlight. Sea water modifies much of the remaining light, reflecting, scattering or converting it into heat. The longest wavelengths of light disappear in shallow sea water; infrared light disappears in the first three feet of water. Red and purple light vanish 30 feet down. Blue-green light penetrates deepest; in very clear water, traces of blue-green light may reach 1,700 feet deep.

Turbid sea water has an entirely different color and personality. Tiny bits of soil or other particles scatter light energy, changing its downward direction and sending light back toward the surface or off at an angle. The changed light gives muddy sea water its greenish hue, and doesn't provide enough energy to satisfy the growing kelps.

From the plant's perspective, not all light energy is equally useful. Land plants with their green pigments (chlorophylls) find red and blue-violet light to be the star performers in photosynthesis. Kelps are more versatile plants; they can take advantage of almost all wavelengths of light. Their chlorophylls harvest red and blue-violet light, and their golden accessory pigment (fucoxanthin) takes advantage of green light. During reproduction, it's blue light that stimulates the kelp gametophytes to make their gametes. In the clearest ocean waters, the most useful colors filter down to the plants and the kelps can get on with their work of photosynthesis and growth.

Streams of sunlight filter through the giant kelp to the forest understory. One of the fastest growing plants in the world, giant kelp can grow 14 inches a day.

SHAPED BY WATER'S FLOW

For better or for worse, each kelp is married to and shaped by the sea. Water moving over the plant brings it dissolved gases and nutrients. Currents disperse the kelp's sperm and spores. Yet violent water motion can be a destructive force, tearing blades, breaking stipes and dislodging holdfasts.

Intertidal kelps face breaking waves, rapid surges and backwash, while offshore kelps sway in swells and tidal currents. Kelps with firm, upright stipes confront the sea with steadfast strength, but other kelps offer less resistance: they flap and flutter in the water flow. Their elastic stipes stretch out in the flowing water and flexibly alternate directions with the surge. Some bull kelp grow coiled stipes that extend their springy play.

Water flow also affects the shape of kelp blades. As it grows, a blade may change to suit its surroundings. The feather boa kelp that starts life soft and sheer stays that way in quiet water, but gets tough and sturdy in rougher waters.

Some kelps have slick blades with slimy surfaces that cut the water's drag or discourage other seaweeds from settling down on them. Other kelps have ribbed and corrugated blades. The ribs and corrugations give the blade strength, while wrinkles, ruffles, holes and spines enhance water flow close to the blade's surface for a greater exchange of nutrients. By looking closely at the beautiful patterns and textures of a blade of kelp, you can learn a little of its life story.

Corrugated channels crease the surface of this giant kelp blade.

Thickened ribs and wrinkles run down the length of a Costaria blade.

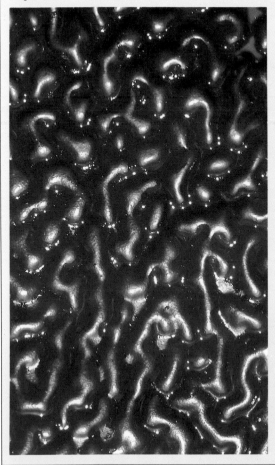

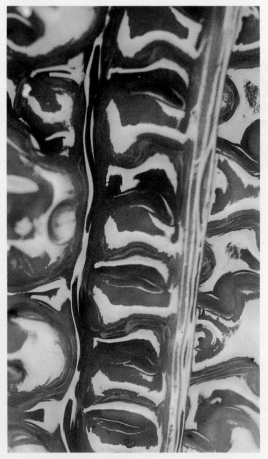

Moderate waves and swells stimulate the kelp forests, while powerful storm waves are very destructive.

WATER MOTION   The ocean has a stockpile of nutrients, but they're not all evenly distributed. Some areas are dilute; other spots are pockets of wealth. Upwelling brings nutrient-rich waters from deeper layers up to the surface in some areas. It's the water motion from currents and wind-driven waves that helps even out the local score and redistribute the wealth.

Even if the surrounding seas have enough nutrients, rapidly growing plants in calm water will occasionally deplete the riches in the thin water layers immediately around them. The kelps may exhaust the very provisions they need most. But, even in the calmest weather, the kelp blade's notched edges and corrugated surface help stir up a little commotion. As turbulence spills across the surface of the plant, it restores the supply of nutrients in the water layers that closely wrap around the kelp.

On a larger scale, the whole kelp community depends on water movement to keep things lively. Like indecisive rivers that change direction, waves and currents push water through the kelp forest, first one way, then back again. The waters flow slowly

at first, a few inches a second, then accelerate where flow is
restricted. As they percolate through the forest, currents alter
the mood, changing temperatures and ferrying in new visitors.
Pulses of living plankton and decaying matter provide food for
the forest animals; larvae and spores settle down to join the
local populations.

Moderate waves benefit the kelp community; forceful ones
are destructive. Rapidly flowing water drags at attached plants and
animals. Storm waves strip old blades and fronds off the kelp and
tear holdfasts from the bottom. Swells remove seaweeds from the
forest, disturbing the neighborhood landscape. Surge and waves
change the bottom, redesigning sand patterns, moving boulders and
fracturing the structure of the rocky reef.

In winter, storm waves six feet tall often charge across Monterey
Bay. Every 10 to 20 years or so, really enormous waves develop
along our coast, reaching heights of 20 feet and moving at
incredible speeds. These gigantic waves assault kelp forests and
their communities before pounding the nearby shore.

3

INTERACTIONS IN THE KELP FOREST

Some folks can't see the forest for the trees, but a kelp forest embodies much more than the golden kelp. Like forests on land, these undersea forests are complex communities with many kinds of plants and animals. Kelp plants play a critical role as a framework for the community. Their three-dimensional structure influences what goes on around them and what grows on and under them.

On the land or in the sea, the most complex forests have a layered or stratified organization. In a kelp forest, the water depth, rocky outcrops, sandy patches and seaweeds of various heights establish the structural layers. Tall or short, deep or shallow, each plant's requirements for sun and tolerance of shade determines where it fits in the hierarchy. Giant kelp and bull kelp need sunny spots. Once established as the uppermost layer of the forest, their canopies shade the plants below. In some dimly lit understories, winged kelp (*Pterygophora californica*) and other shrublike kelps stand nearly as tall as people. Seaweeds at lower levels get less and less light in turn, as each layer absorbs and deflects the sunlight that reaches it. The multiple layers in a kelp forest offer many different kinds of habitats. Often, the variety of different habitats promotes a greater diversity of inhabitants.

In the kelp forest, plants, invertebrates and vertebrates collect and interact with one another. Occasionally, the interactions are neutral; sometimes they're beneficial and sometimes hostile. Together, all the creatures and their interactions form a dynamic, living organization: the kelp forest community.

Every kelp forest creature confronts the same challenges that face living creatures everywhere. Where can it live? What can it eat? Will it survive long enough to reproduce?

These understory kelps grow six feet tall in the dim light filtering through the giant kelp canopy (left).

At the base of a giant kelp, a cluster of repro-ductive blades flutter atop a well-developed holdfast (right).

Kelp forest plants and animals, like other organisms, adapt to their particular surroundings. When change occurs in the kelp forest, it alters lives and lifestyles of the forest inhabitants. Adjustments arise even under water: there are shifts in the weather, competition for mates or space, predators on the prowl and disease epidemics. Change is an essential ingredient of life in the forest.

PHYSICAL DISTURBANCE Kelp forests thrive in quiet, protected bays and along wild, wave-swept coasts, but the structure of each forest shifts to suit the site. Plants like the bull kelp seem specially built to handle the stress of wind and waves. When waves and currents tug at the bull kelp's single float, its stipe stretches out to go with the flow. Plants like the giant kelp stretch out too, but are more vulnerable because their multiple fronds, floats and blades add more drag on the holdfast. Each kelp survives where it can: giant kelp dominates the protected waters of Monterey Bay, while bull kelp reigns in the rougher waters outside the bay.

REPRODUCTION Reproduction is like a life insurance policy that promises a particular kind of organism will persist beyond any one individual's lifespan. But it's a promise with no guarantees. From the demise of the dinosaurs to the passing of the passenger pigeons, plenty of plants and animals have already disappeared to extinction.

Some kelp forest creatures are locked into a yearly reproduction cycle. For others, each year brings a new story. Annual plants like the bull kelp put their energy into that one year's reproductive effort, a one-time burst of spores. By the time the spores settle down and develop into the gametophytes that reproduce sexually, the original bull kelp is long gone, leaving room for a new tenant on the scene.

On the other hand, perennials, like the giant kelp, can be a hard act to follow. Once established, a giant kelp's likely to persist for two to seven years, taking up space and blocking out sunlight. In the shadow of such a persistent parent, a giant kelp spore has little chance of making its own place in the world.

Summer and winter, night and day, there's almost always some reproductive activity underway in the kelp forest. With sexual encounters or asexual arrangements, the plants and animals lay the groundwork for future generations. Some creatures keep it simple: they reproduce by dividing in half. The beautiful diatoms of the kelp forest plankton use this asexual method. A diatom is a microscopic plant, a golden-brown cell enclosed in a glass-walled box. The box is made of two slightly unequal parts: a larger glass bowl fits over a smaller one to form a protective pillbox. As the diatom swells in size, it pushes its two sections apart, makes new bowls on the inside and then divides into two diatoms. One diatom is the same size as the original, and one is slightly smaller, but each has all the essential parts.

Budding is another kind of reproduction that doesn't require male-female encounters. The bryozoan (*Membranipora tuberculata*)

Like anchored cables securing the floats and blades, these fronds of giant kelp bend in the pull of waves and swells.

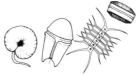

A solitary diatom pillbox floats over a spiked chain of diatoms and two rounded dinoflagellates. These microscopic drifting plants called phytoplankton form the first link in many ocean food chains.

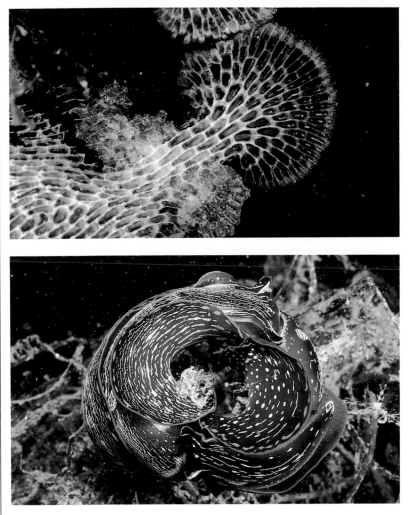

Budding off its oblong cells in microscopic rows, this bryozoan colony fans out across a blade of kelp.

Sea slugs, like this pair, are hermaphrodites: each one has both male and female sexual organs. Joined head to tail in mating, these sea slugs fertilize each other's eggs.

is a small invertebrate that expands from a party of one to a crust of thousands. The individual bryozoan is tiny, perhaps as tall as this word (1/32 of an inch), but a colony can encrust an entire kelp blade. The initial bryozoan larva buds off little bumps. These clones remain fused to their original buddy as they develop from subordinates into supporters. United in a common cause, the bryozoans work together, each sprouting new buds to fan out into a colonial empire.

Sex complicates life for many plants and animals; it adds the problem of finding a mate to the rest of life's challenges. To its credit, sex offers a reshuffling of genes that blends the parents' traits into new combinations in the offspring. Snails, clams and vertebrates, like mammals and fishes, rely exclusively on sexual reproduction despite the problems of getting together with members of the opposite sex. Many organisms take the best of both worlds, fragmenting or cloning some of the time, while maintaining sex as an option when the time is right.

Some kelp forest creatures are hermaphrodites, with both male and female sexual organs. A few hermaphrodites can fertilize their own eggs. Others, like the showy sea slugs—the nudibranchs—can maximize each sexual encounter. Any two nudibranchs can mate, and when they do, twice as many eggs are fertilized at one time.

Many kelp forest invertebrates, like the sea stars, as well as some fishes, put their reproductive energy into producing huge numbers of eggs and sperm that they disperse directly into the water. The release of eggs by one female often triggers spawning in other nearby males and females. Eggs are fertilized right out in the open and the young develop into larvae as they ride the currents. They're independent drifters from the moment of fertilization.

Rockfishes (*Sebastes* spp.) follow a different plan: males fertilize the eggs inside the females that later give birth to large numbers of live young. The newborn rockfishes are tiny and quite immature when they're first released, but they continue to develop as they drift with the plankton.

Marine mammals and some fishes, like the kelp surfperch (*Brachyistius frenatus*), produce fewer babies that are larger when they're born. Their larger size may give the newborns a better chance to survive. Kelp surfperch emerge ready to swim and feed in their parents' neighborhood.

Zooplankton—tiny animals like these microscopic larvae and the tiny buglike copepod (above)—drift through the kelp forest, providing food for filter-feeding animals.

Two bat stars touch to skirmish, but not to mate (left). Like other sea stars, they release their eggs and sperm directly into the water.

Female surfperches give birth to their young in litters. This well-developed newborn is emerging tail-first from its mother.

DISPERSAL There's always the question of where to put the offspring. Is it safer to keep them close to home or worth the risk of sending them off to explore on their own?

Almost all kelp forest creatures, from mobile crabs and snails to stationary seaweeds and barnacles, spend some part of their lives drifting in the water. A free-roaming stage offers a chance of finding new, unclaimed territory to colonize, although in reality, most of the drifters are eaten or lost at sea.

BIOLOGICAL RELATIONSHIPS When plants and animals share space, they interact with one another and often influence each other's survival. Some of these relationships seem neutral, others are positive or negative. Mating is one type of positive partnership that can be gratifying for the individuals as well as for the species.

Symbiotic relationships between different kinds of organisms can also be positive: living together improves the odds that one or both will survive. Some symbiotic relationships are restrictive but dependable; the partners are obligated to stick together. Other partnerships aim for a more flexible convenience; separation and divorce remain an option if things don't work out.

Some associations benefit only one of the participants, while the other is neither harmed nor helped, like the symbioses of certain fishes and anemones. In the kelp forest, strawberry anemones (*Corynactis californica*) stand less than an inch tall in colorful clusters on rocks. Some colonies are red as strawberries; others are pink, purple or yellow with white-tipped tentacles that bear poison nematocysts (pronounced "knee-**mat**-o-sists"). The nematocysts carry tiny darts that explode on contact to stun small prey and discourage hungry predators. A few fortunate fishes, like the painted greenling (*Oxylebius pictus*), wriggle up and down among the tentacles to take advantage of the anemones' stinging defense. The fish protects itself, perhaps with a coat of slime and its dainty dance between the tentacles, so it can nestle unharmed among the anemones. Safe from predators that fear the anemones'

Against a colorful background of sponges, an anemone swallows a huge bat star meal (below left).

Immune from the anemones' stinging cells, a painted greenling rests among strawberry anemones (below right).

sting, the painted greenling may rest at night in a bed of the strawberry anemones.

In some symbiotic relationships, different organisms live together with benefits for both participants. On shallow kelp forest floors, large aggregating anemones (*Anthopleura elegantissima*) stand on elegant green or white columns. The green ones get their color from microscopic algae that live inside their tissues. Protected inside the anemone, the algae somehow direct the anemone to expand in sunlight and contract when the light gets too strong. The algae use sunlight and the animal's wastes for photosynthesis, then send back food in return. In deeply shaded habitats, the anemone loses its algae and becomes a pale, slower-growing ghost, indifferent to light and dark.

A clone of identical aggregating anemones crowds together on a rock. Microscopic algae inside the anemones contribute to their greenish color.

COMPETITION Competition is a way of life in the kelp forest as groups struggle over limited resources like food or space. Seaweeds compete for the sunny spots. Fishes fight for the available food. An abundance of one kind of prey spurs some animals to become diet specialists, concentrating only on that food.

Rivalry between kelp forest inhabitants is at times indirect and subtle, or involves an open physical struggle. Sometimes representatives of the same species clash; other times opponents aren't related at all. Usually, the more alike the two competitors are, the more intense the competition.

Two of the many kinds of rockfishes inhabiting the kelp forest compete for food during some part of their lives. The black-and-yellow rockfish (*Sebastes chrysomelas*) and the gopher rockfish (*Sebastes carnatus*) both live in holes and crevices in the rocky forest floor, and both favor crabs, shrimps and octopuses for food. They're so similar in appearance, it's nearly impossible to tell them apart except by color. To relieve the rivalry among these rockfishes, each fish sets up a territory and guards it against intruders. Most black-and-yellow rockfish stake their claims on the shallower reefs; gopher rockfish generally settle in deeper water.

PREDATION What could be more important than food? Plants make their own through photosynthesis, so for them, sunshine and nutrients in the water are the necessities of life. For animals, securing food is crucial. Whether they're herbivores or carnivores or favor a mixed diet of plants and animals, the hunt for food dominates the lives of most kelp forest animals.

On the rocky forest floor, red and purple sea urchins (*Strongylocentrotus franciscanus* and *S. purpuratus*) feed on kelp and other seaweeds. Many snag drift kelp to eat, but when drift is scarce, urchins roam the rocks, cropping off living seaweeds. Unchecked, a horde of hungry urchins will eat everything in reach, leaving barren rocks where a kelp forest once stood. Sea otters (*Enhydra lutris*) and sunflower stars (*Pycnopodia helianthoides*) eat sea urchins, helping to limit the number of urchins. This kind of population control preserves the forest by reducing the number of urchin competitors in the kelp-eating contest.

PARASITES No creature is completely safe from parasites, which are often smaller than their prey. All the kelp forest creatures, from the seaweeds, invertebrates and fishes to occasional visitors like gray whales (*Eschrichtius robustus*), face a struggle against fungi, bacteria, viruses and other parasites. Some parasites are long-lost relatives that take advantage of their hosts, like the small seaweeds that have lost their photosynthetic skills and live on seaweed relatives, stealing their sugars and other vital nutrients.

Crustaceans, like the kelp crab (*Pugettia producta*), sometimes act as hosts to other crustaceans, like the barnacle parasite (*Heterosaccus californicus*). The immature parasite looks much like other barnacle larvae afloat in the water, but it's not as independent as the usual kind. This barnacle just won't grow up without its crab host. When it locates a kelp crab, the barnacle grabs on and impales the crab with an antenna. Shedding most of its body parts, the barnacle slides through its own hollow antenna into the crab's body. Once inside, the barnacle becomes a ruthless invader, using the kelp crab's body to breed more barnacle parasites.

The black-and-yellow rockfish lives in crevices in the rocky forest floor (top left). Sometimes, a related rival—the gopher rockfish—(top right) competes with it for food or habitat.

Red and purple sea urchins cluster around a solitary kelp stipe as they scour for food in a denuded kelp forest (above left).

Lying on its back on the water's surface, a sea otter devours its sea urchin meal (above right).

FORESTS BY LAND OR BY SEA

Sunlight streams through a redwood forest along the coast of California.

The golden canopy of a kelp forest in Monterey Bay filters sunlight, shading plants below.

Our familiar terrestrial forests share a common bond with marine forests. They're both communities that support many layers of life and experience seasonal changes. Plants provide the structure, from tall canopy species to underbrush, and animals in these communities must find niches in the multilayered framework.

On land, tree roots penetrate deep into the ground to get the water and nutrients they need. The forest trees mature slowly. As they grow, they form canopies that shade the understory. Leaves and branches fall to the forest floor, fortifying the soil with nutrients. Rich forest soils support millions of microorganisms: algae, worms, mites and tiny insects. Large animals, like white-tailed deer, graze in the understory, while racoons, squirrels, bears and mice subsist on seeds and hunt for varied foods. Predators—snakes on the ground and hawks in the air—hunt for prey. Scavengers, like vultures, and earthworms break down dead matter.

In forests in the oceans, kelps and other seaweeds draw their nutrients directly from the seawater rather than through roots. When nutrients are abundant, kelps grow quickly, but compared to long-lived redwood and oak trees, kelps live short and sweet lives lasting one to seven years. Most never get a chance to die or decompose in place. Torn free by storm waves, they drift out of the forest, although urchins and abalones may snag and eat some of the drifting kelp. In turn, these urchins, snails and other seaweed-eaters provide food for the carnivores in the kelp forest. Sea stars slowly swallow snails and fishes search for crawling prey. Sharks and sea lions hunt the hunters. Versatile crabs scavenge for animal remains or pursue living prey, while sea cucumbers feed on bacteria as they vacuum sediments off the forest floor.

Whether on land or under water, plants provide habitats for forest animals and ultimately support all forest food webs.

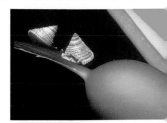

DEFENSES Although grazers and predators hunt in the kelp forest, their prey are not entirely defenseless. Plants and animals have an impressive array of adaptations that help them avoid being eaten. Some creatures defend themselves passively with spines, shells, tough textures and poisons; others hide, run or fight.

Leaving the scene of the crime can be a good way to avoid being eaten. Even slow-moving snails sneak away if they can outpace slower predators. The channeled top snail (*Calliostoma canaliculatum*) is sensitive to sea star scents. If it detects the scent of a predatory sunflower star, the snail "races" away or climbs up a kelp. Sometimes it's quick enough to escape; sometimes the sunflower star's quicker. The star's caress provokes the final frenzy: the snail emits a noxious yellow slime, then lifts its shell and somersaults away, even tumbling off a ledge to get away.

Purple sea urchins can't run away fast, but they maintain a strong defense with shell and spines, pinchers and noxious chemicals. Urchins sometimes drill strongholds in the rocks. Using their grinding teeth and spines, they slowly erode homey holes and depressions in boulders. If an urchin outgrows its exit, the fortress that keeps otters out becomes a homemade prison that keeps the urchin in. Safe, but segregated, the prisoner's forced to feed on whatever seaweeds drift within reach.

Camouflage is another way to hide—by blending into the environment. In kelp holdfasts and on rocky reefs, the decorator crab (*Loxorhynchus crispatus*) is a master of disguise. Even out in the open, it can hide in plain sight. Decorations on the crab's shell provide its camouflage. This small crab collects bits of seaweed, hydroids and sponges and attaches them to hooks on its back and legs. As the crab's back-garden grows, its camouflage improves. Against a background of sponges, the sponge-covered crab is indiscernible. Amid a patch of red algae, the seaweed-shrouded crab is just another bump on the horizon. Moving slowly, the well-decorated crab can bypass hungry, hunting fishes and sea otters.

At times, the best defense against attack is a spirited offense, whether it's a physical or chemical attack. The gaudy colors of many sea slugs advertise that these nudibranchs carry potent weapons. Their skin secretes acid or poison, or they'll protect themselves with the stinging cells of anemones they've eaten. Some, like the brilliant orange-and-blue *Hermissenda crassicornis*, are aggressive fighters. When two of them meet head-to-head, they'll mate or lunge into a biting battle. If one *Hermissenda* meets the tail of another and gets in the first lick, it's likely to win the war and consume the loser.

Waves and weather, competitors and predators all influence life in the kelp forest. In each part of the kelp forest, residents compete for space, mate, cohabitate, eat and are eaten.

Two ringed top snails graze on a giant kelp plant (top).

A purple sea urchin nestles in a rocky crevice (second from top).

A background of orange sponges reveals the decorator crab in its bryozoan camouflage (third from top).

This colorful sea slug defends itself with stinging cells reserved from its hydroid meals (above).

4

KELP FOREST COMMUNITIES

Like small towns, biological communities spread out across a variety of neighborhoods, some safer, more attractive or more convenient than others. Within any neighborhood, there's an assortment of individual homes, each with its distinct architecture and landscaping. As home owners everywhere know, it's not just the structures, it's neighbors that make the neighborhood.

In the kelp forest, kelp plants and the seafloor form the neighborhood subdivisions: the kelp canopy, stipes and holdfasts, the water, reef tops, vertical faces, crevices and sand channels. Each neighborhood is unique; contours, inhabitants, even the atmosphere changes from habitat to habitat and from forest to forest. Some residents are versatile; they'll survive in any number of homes. A few are specialists; if they can't find the perfect spot, they won't settle down. The best properties are taken first. It's a lucky resident that finds the perfect vacant address to suit its lifestyle, but there's always room for some adjustment in the community. Disturbances open up spaces: waves tear through, sands scour the rocks, predators scrape off food. Drifting spores and larvae in the water will settle out when there's an opening for new recruits.

Giant kelp influences everything that lives in the forest. The plants stand like a windbreak to rolling Pacific waves, and their close-spaced fronds help screen animals from hungry predators. In the forest there are dwellings available at every height and depth. Some kelps ascend from rocks 12 to 15 feet below the surface; others rise to the surface from 60 to 100 feet down. Different depths shelter distinctive communities. On the shallow sunlit rocks of the forest, seaweeds dominate the cast of characters, while invertebrates command the deeper, darkened recesses. Most seaweeds seek the firm support of solid rocks. The sandy channels offer the plants no foothold, but many animals find shelter there.

IN THE KELP CANOPY On the sunlit surface of the kelp forest, a few blades of kelp stretch out to catch the best light. These vital surface blades crown the older fronds. They transport sugars back down the stipe to support the holdfast and younger fronds below. Beneath the surface, younger fronds quickly grow upward to thicken the canopy, providing layers of shelter and food for animals in the neighborhood. Fishes hide out between the blades. The tiny juvenile stages of invertebrates settle out of the plankton onto the blades and metamorphose into adults. Eventually, thousands of animals occupy secret spaces in the canopy.

Outside the safety of their holdfast homes, colorful brittle stars writhe their long, flexible arms (above).
In kelp-colored disguise, a giant kelpfish peeks out of the canopy (right).

Floating larval stages of hydroids (*Eucopella* spp.) and bryozoans drift through the canopy. A hydroid larva hovers, then settles on a blade of kelp. Securing itself in place there, this tiny animal perches like a stalked flower. The hydroid divides to spread its delicate fuzz into a microscopic garden across the blade. The stalks grow a bit taller, then branch to form a network of tubes, each crowned with a dainty cup. The hydroid colony poses as a peaceful garden, but it's a military camp ready for combat. Each flowerlike cup is armed with stinging cells to capture microscopic plankton. Inside the network, the tubes interconnect, so when one part eats, the whole colony shares the meal. The stumps of timeworn hydroids cover some of the older kelp blades. Once tall and graceful, these hydroids were grazed to stubs by crabs and sea slugs.

A school of small, transparent opossum shrimp (*Acanthomysis* sp.) has established a cloister in the canopy, hidden between the blades of kelp from predatory squids and fishes. (Like their namesakes the opossums, they carry their young in pouches.) By day, these delicate shrimp stay in seclusion in the kelp. At night, they venture out to snatch up smaller crustaceans.

Some fishes hover in hiding near the kelp canopy. A two-foot-long giant kelpfish (*Heterostichus rostratus*) sways in perfect synchrony with the forest. A golden kelp-colored brown, it masquerades as a kelp blade. Far below, another giant kelpfish wears a different color combination: it's a darker brownish-purple with stripes. Why? Perhaps the dark one's better camouflaged against the forest floor. Both kelpfish match the languor of the water movement and slowly scan the water for crustaceans and small fishes. Nearby, a reddish male kelpfish protects a clump of seaweed that bears his future offspring. A female laid her greenish eggs on the seaweed; now her blush-colored mate stays close at hand to guard them until they hatch.

Like microscopic flowers, a fuzz of hydroids spreads across the kelp blade (above left).

By day, these delicate opossum shrimp conceal themselves between blades of giant kelp (above right).

ON THE KELP STIPES Below the surface, columns of kelp rise like anchor lines from the seafloor to the canopy. In this kelp column neighborhood, homes are a mellow mixture of narrow and thick vertical cables. Older, elongated fronds blend with juvenile blades growing upward toward the canopy. This harmonious neighborhood isn't exclusive; the spider crabs and rockfishes living here would be equally at home in the canopy or near the bottom.

Large turban snails glide slowly up and down the stipes. Three kinds of turban snails share space in the kelp forest: the brown turban snail (*Tegula brunnea*), the dusky turban snail (*T. pulligo*), and the Monterey turban snail (*T. montereyi*). The uncommon Monterey turbans seclude themselves in deeper waters, but dozens of the brown and dusky turban snails hustle to make their mark on the stipes. The snails cling firmly at lofty levels through all but the strongest storms and swells. With hungry sunflower stars, sea stars (*Pisaster* spp.) and rock crabs (*Cancer antennarius*) lingering on the kelp forest floor, most snails seek supper and security at higher positions. Grazing on the stipes as they move up and out of danger, the turbans leave rake-marks on the succulent fronds.

Slender fishes stake out sections of the stipes. A kelp-colored clingfish (*Rimicola muscarum*) clings to the kelp column, moving only to take cover or seek food. Like a two-inch tadpole with a large head and tapering body, this kelp clingfish isn't built for speed or show; it's built to cling. Suction from an adhesive disk on its underside holds the clingfish tight against the kelp where it hides from larger fishes.

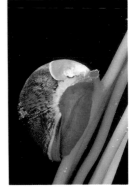

A slender clingfish secures itself to a kelp stipe (above left).

By day, this southern California top snail wanders down the kelp stipes. At dusk, it heads back up (above).

Half hidden in a mat of crunchy coralline algae, the rock crab scans the forest floor for snails (left).

IN THE WATER The watery passageways between the kelp columns create a city of canals for swimming fishes and drifting plankton. The tiny drifting plants or phytoplankton change with the seasons: jewel-like diatoms in spring and dancing dinoflagellates in fall.

A zoo of animal plankton mixes with the drifting plants: copepods and shrimps, larval crabs and fishes. Many hang low in the water by day and rise up to the surface with darkness. Some of these tiny animals recently hatched from eggs and still sustain themselves on the egg yolks they're carrying with them. Others nibble on microscopic plants and other tidbits. Fishes find this drifting zoo and nibble on the plankton.

Loose schools of yellow, cigar-shaped señoritas (*Oxyjulis californica*) swarm through the forest. Buck teeth protrude from the small mouths of these picky eaters, ideal for picking bryozoans and hydroids off kelp and copepods off other fishes. Disturbed by potential predators, señoritas dart to the seafloor and hide by burrowing in the sand. At night, they rest there too, buried in the sand with just their heads exposed.

The cigar-shaped señorita darts through the canopy to pick small invertebrates off kelp blades and other fishes.

INSIDE THE KELP HOLDFAST At the bottom of the kelp forest, holdfasts anchor the kelp plants in place. Between the knotty branches of the kelp holdfast, many shy and secretive animals find a protected neighborhood. Bits of decaying food that collect in the small holdfast attract the first settlers. As the tangled holdfast grows, more and more residents seek refuge there. Sheltered from large predators and weather, the animals feed on plankton, kelp, debris or other residents of the holdfast.

At night, some kelp holdfasts seem to sprout whiskers when hundreds of colorful brittle stars (*Ophiothrix spiculata*) thrust fuzzy arms from between the branches. Five long arms surrounding a flat, central disk mark the brittle stars as distant relatives of the sea stars. With one long arm hooked inside the holdfast, each brittle star extends its remaining arms outside to catch bits of food. Using its tube feet as a conveyor belt, the brittle star transports tasty particles from sticky spines on its arms to its mouth on the central disk.

Multitudes of worms wriggle through the holdfast. Segmented worms, the mobile polychaetes wander through the tangle in search of food. Bristles project from each segment of their sectioned bodies. The segments contract and then expand to propel the worms forward. Their many limbs are multipurpose organs, collecting oxygen, providing mobility and sensing the surroundings as they go. Some worms are carnivores with formidable jaws and teeth that make quick meals of smaller organisms. A few use fleshy lips to scoop up nutritious sediments.

Carrying an abandoned turban snail's shell on its back, a hermit crab (*Pagurus samuelis*) scavenges for kelp and dead animals on the outer fringes of the holdfast. This crustacean's asymmetric abdomen curls up snug inside the protective shell. A hermit crab won't bother shells still occupied by living snails, but it'll fight to steal one off the back of another crab.

The hermit crab protects its soft body parts inside an empty snail shell (above). When the crab outgrows its shell, it finds a new one.

While large sea stars linger on the outside, hundreds of secretive animals find shelter deep inside the holdfast (left).

On the edge of the holdfast, two hermit crabs meet, not as rivals but as mates. In a courtship dance, the male crab grips the female's shell and carries her about, pausing from time to time to knock his shell against hers. He courts and carries her all day, until the female crab lowers her guard and extends her soft body from the shell's protective armor to quickly mate. Abruptly, they separate and retreat to opposite ends of the holdfast. Inside her shell, the female hermit crab carries a new burden: over 2,000 fertilized eggs.

Hundreds of other animal feet march through the holdfast: copepods, amphipods, isopods and decapods. Flatworms and roundworms, snails and slugs slink, slither or secure themselves among the branches. In tight quarters, clams and mussels flex their muscles for a bit of space.

FOREST FISHES

More than 150 different kinds of fishes live in the coastal waters off California. Few of these fishes dwell exclusively in kelp forests, but the variety of forest habitats offers options for different kinds of fishes. You can tell where a fish fits in the forest by a few clues, like its shape, color and habits.

Sleek and streamlined swimmers, Pacific sardines are constantly on the move. Traveling in schools, they feed on plankton. Silvery flashes in their midwater habitat, they leave slower fishes behind. Forked tails give them speed and power, and their fins are evenly distributed for stability and maneuverability.

Lie-in-wait predators like the cabezon and other sculpins have camouflaged bodies patterned to match their seaweed hiding places. Fins set back on the body and a large tail give the cabezon the extra thrust it needs to launch surprise attacks. Its upward-pointing mouth opens wide to snap up crabs and fishes.

Midwater fishes are often neutrally buoyant—hovering with ease in midwater. Some fishes fight the pull of gravity with low-density oils in their bodies; some have shaped and angled swimming fins that generate lift. Other fishes, like rockfishes, have a swimbladder, a gas-filled space that helps regulate their buoyancy.

Fishes like the tubesnout change their habits and their habitats. The tubesnout is covered with bony plates so it's rigid and can't turn quickly. It swims slowly near the water's surface, camouflaged from below by its light-colored underside. After mating, the male tubesnout guards the eggs. Hidden by its kelp-colored back, the male hovers near its nest of kelp.

Many bottom-dwelling fishes, like rays and flatfishes, are flattened for continuous contact with the seafloor. These fishes usually lack swim bladders so the negative buoyancy that makes them sink to the bottom works to their advantage.

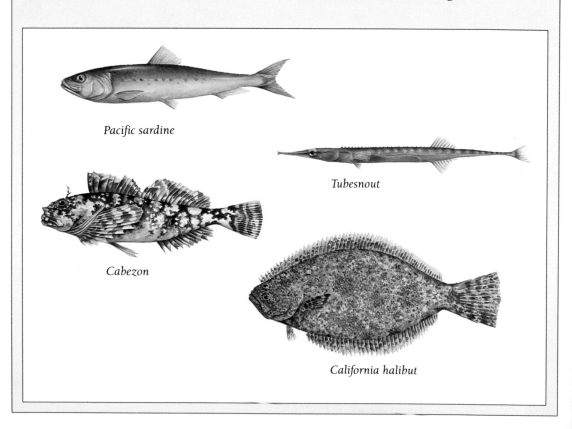

Pacific sardine

Tubesnout

Cabezon

California halibut

UP ON THE REEF TOPS Most people think "reefs" mean coral reefs, but a reef is any hard structure in the sea. Old cars, concrete blocks and sunken boats that provide sites where underwater life can attach or hide are called "artificial reefs." In Monterey Bay, slabs of sandstone and chunks of granite form natural reefs where kelps can attach. The hard, horizontal surfaces of the reef tops are speckled with sunlight that fades and brightens as the canopy sways overhead. It's a perfect rock garden for seaweeds in the flickering sun and shade.

Ranked by seniority and stature, understory plants compete for positions in the dwindling light. Between the holdfasts of giant kelp, smaller shrublike kelps stand nearly as tall as people. Winged kelp extend their armlike blades in welcome, while *Eisenia arborea* swing twin pony-tail blades like schoolgirls.

Lovely herblike seaweeds flutter in the dim light below these kelps. The delicately branched brown *Desmarestia ligulata* has an attractive facade, but an acid disposition. Inside each cell, it conceals a tiny bomb of sulfuric acid to discourage grazing fishes. Crunchy patches of pink coralline algae (*Calliarthron* spp. and *Bossiella* spp.) flourish on the reef top. Like bony hands, they wag discouraging fingers at passing fishes; these calcified plants are a hard meal to swallow. Flexible knuckles alternate with the hard calcified parts so these plants can bend with the waves.

A perennial turf of fleshy red algae carpets the reef tops: narrow tongues of red *Rhodymenia californica*, clumps of sea grapes, bumpy Turkish towel (*Gigartina corymbifera*) and iridescent *Fryeella gardneri* glowing blue. Ephemeral delights come in all shapes: golden featherlike plants (*Giffordia* spp.) quickly colonize bare space on the reef; relaxed balloons of *Coilodesme californica* appear on the taller spikes of brown *Cystoseira osmundacea*.

Understory kelps and pink coralline algae compete with the invertebrates for a spot in the speckled sunlight on the reef top (above).

The joints in these hard corallines give flexibility.

Robust Rhodymenia stands upright.

Bumpy Turkish towels flap in flowing water.

Snails, small crustaceans and worms wiggle through the tangle of weeds. Colorful sea slugs flaunt their showy gills. A blaze of stars, red, orange, pink and purple, brightens the seafloor. One giant sea star (*Pisaster giganteus*) stretches out its five blue arms in search of food and locates a patch of barnacles (*Balanus* sp.). Attached to a rock, the barnacles close up tight, but their shell fortress is no match for the persistent pull of the sea star's tube feet. The sea star tears a small patch of barnacles from the rock and begins to feed.

Excited by the scent of food, a huge sunflower star advances on the picnic. It shoves its twenty rays against the smaller sea star, but this blue *Pisaster* is no shrinking violet. Armed with many small seizing organs, it responds with a pinching counterattack to the sunflower star's sensitive underside. The injured sunflower star withdraws its challenge and *Pisaster* returns to its meal.

Many fishes of the reef floor lie camouflaged in the seaweeds, motionless until a meal comes along. Others cruise by on exploratory surveys. These low-life fishes aren't social climbers; they don't have swim bladders, those special organs that help other fishes hang at higher levels.

A large secretive sculpin, the well-camouflaged cabezon (*Scorpaenichthys marmoratus*) changes colors to match its surroundings as it pokes along the reef top. Spying a small red abalone, the fish bumps hard against it to dislodge it from the rock. The cabezon swallows the abalone whole, and later regurgitates the shell.

A sedentary lingcod (*Ophiodon elongatus*) lurks in the seaweeds, ready to ambush whatever swims by. The tiny pink crustaceans decorating this greenish fish look pretty, but they're parasites. Waves of these small copepods race over the fish; other times they slow down to nibble bits of skin. When a small rock-fish ventures close, the lingcod gets its turn to feed. Lunging for the smaller fish, it seizes its prey in a quick burst of speed.

An active predator, this giant sea star hunts for mussels, snails and barnacles (top).

The largest sea star of our Pacific coast, the sunflower star uses speed and power to overtake prey (above).

Wearing a crown of gemlike crustaceans, the lingcod provides a home and food for these small pink parasites.

AGAINST VERTICAL WALLS There's a natural high-rise condominium on the steep faces of the reef. Reef walls stretch upward as a vertical complex of homes. Sponges, sea squirts, anemones and other animals form a multilayered community, occupying different depths and positions on the rocks. Slow-growing, attached creatures persist in the same place for decades, giving a sense of continuity to the neighborhood. Other residents time-share their spaces; they're here for a season, and, when they're gone, someone else takes their place.

On the dimly lit vertical rocks, only a few seaweeds survive. Crusts of coralline algae add splashes of pink to the colorful mosaic of sponges, hydrocorals and anemones. Water currents bathe the neighborhood, bringing drifting food to the filter-feeding invertebrates. The flow of water also carries off sediments that might smother the delicate, attached residents here.

Vivid purple and pink hydrocorals (*Allopora californica*) form erect branching colonies of tiny anemonelike animals. Like their hydroid relatives, hydrocorals harbor stinging cells that help capture tiny plankton food. The colony acts as a unit with three kinds of members, each specializing in one kind of work: to feed, defend or reproduce for the benefit of all. On horizontal surfaces, the colonies die young, quickly overgrown by weeds or smothered in silt. On slopes and cliffs, they live long, colorful lives. Distinctly dazzling hydrocorals are jewel-studded—small, white barnacles (*Armatobalanus nefrens*) grow embedded in the colonies and nowhere else.

Bright orange corals (*Balanophyllia elegans*) pose royally on small cuplike thrones, safeguarding secret treasures. The coral's

This carnivorous cowry consumes the strawberry anemones that grace these vertical reef walls (above).

Pink and purple hydrocorals are colonies of anemonelike animals (below).

An orange cup coral retracts into its stony skeleton (bottom).

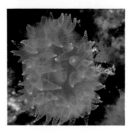

central mouth opens wide, releasing small larvae that were developing inside. The larvae don't travel far from home; they crawl just a foot or so from their parents, then settle down and secrete their own limestone thrones. Once settled, the corals usually hold their ground for six or more years unless a more competitive creature overwhelms them (or they're eaten by spider crabs).

Red, orange and yellow sponges form a colorful patchwork on the reef walls. Sponges are simple animals with spectacular shapes: crusts and cups, balls and branches. Skeletons of spiny spicules support the sponges' bodies. They're riddled with canals that open to the surrounding water. The canals are lined with chambers containing whiplike flagella that work like motors to drive in water currents carrying oxygen and planktonic food, and transport out debris and wastes. Roomy sponge canals make good homes for other creatures, too, like tiny parasites. Sea slugs wandering near the sponges sometimes stop to feed on them. The sponges' noxious taste and prickly spicules don't seem to slow down hungry sea slugs.

IN REEF CRACKS AND CREVICES   Ledges, caves and crevices in the reef provide temporary housing for roving invertebrates and fishes. Other current-washed spots harbor permanent populations of the same filter-feeding residents found on the vertical reef walls.

Hidden in a secluded spot, a peanut worm (*Phascolosoma agassizii*) stretches long and thin, then contracts back into its fat peanut shape. Sedentary housekeepers, these worms lie around in holes or holdfasts and slowly vacuum up bits of debris to feed on the filmy coverings. In spring, the female worms eject eggs into the water; males toss out their sperm to fertilize the drifting eggs. Developing from fertile eggs, the tiny peanut larvae are carried off by currents until they're mature and ready to settle in a suitable crevice.

Sea urchins and abalones wedged in cracks wait for edible pieces of kelp to drift within reach. A red octopus (*Octopus rubescens*) moves by with fluid grace along the rocks. Its large eyes alert for movement, the octopus explores cracks and crevices in the wall with its arms as it passes. Spying a crab, the octopus wrestles with its prey and squeezes into a nearby hole. As resident crabs and shrimps scurry away, the octopus tears into its meal.

Six-foot-long snaky shapes with dark spots that help their camouflage, wolf-eels (*Anarrhichthys ocellatus*) maneuver gracefully in and out of caves. Searching for food, they poke their snouts into cracks and crevices. One grabs a crab in its powerful jaws and devours it shell and all. In deeper water, a mated pair of wolf-eels share a rocky grotto. The female lays her eggs among the rocks and both fish guard them until the young hatch.

Colorful invertebrates cover nearly every inch of available space on the rock walls in the kelp forest (right).

A red octopus can squeeze into the smallest hole.

A wolf-eel maneuvers in and out of rocky crevices.

IN THE SAND CHANNELS Sand and shell fragments accumulate in the furrows between reefs. For tunnel diggers and tube dwellers, these gritty gutters make the best kind of home. A burrowing animal can shape a form-fitting home in these soft sediments and still have room to stretch out and grow. Drifting seaweed and other bits of food sustain the animals that live here.

Stationary segmented worms command a leading role in these sandy neighborhoods. Some, like the ornate tube worms (*Diopatra ornata*), are solidifying citizens. They manufacture sticky parchment tubes that hold the sand in place and stabilize the neighborhood. Moving gently up and down, these worms pump water through their tube homes. The homemade currents bring in odors from the neighborhood so the worms know what foods are available. At night, they poke their heads out and wait for food to drift by. With their powerful jaws, these tube-dwellers catch and eat small invertebrates and bits of drift kelp while other worms share the rain of smaller food particles.

A burrowing anemone (*Pachycerianthus fimbriatus*) extends its graceful tentacles into the water. Stinging nematocysts in the streaming tentacles immobilize small, drifting animals. The tentacles carry captive prey to the anemone's mouth. The anemone's halo of tentacles lures a hungry sea slug, the rainbow nudibranch (*Dendronotus iris*). Creeping up the anemone's tube, the sea slug stretches and chomps onto a tentacle. In defense, the anemone immediately contracts and withdraws into its tube, dragging the sea slug inside. Confined in the anemone's tube, the sea slug finishes its tentacle meal, then crawls out and away, leaving the anemone to recuperate.

Orange and red, plain and mottled, a colorful spectrum of bat stars (*Patiria miniata*) stretches out on the sand. These stars eat just about anything: plants or animals, dead or alive. Odors of decaying fish will launch a slow-motion bat star feeding frenzy. Moving together toward a carcass, they push and shove in gentle combat. Piling up on the fish, the bat stars extend their stomachs to engulf their rotting meal.

Gumboot chitons (*Cryptochiton stelleri*) cruise slowly across the sand into a pile of drift seaweeds. With toothed tongues, they ingest the red leathery blades, rasping in the food. Nearby, a female gumboot chiton lays her eggs in gelatinous spiral strings. Intent on her work, she won't stop to eat, even though she'll lose weight in the process. Her eggs trigger nearby males to release sperm. In less than a week, tiny chiton larvae develop in the gel and drift away to settle elsewhere.

The eerie eyes of rays and flatfishes peer up from the sandy bottom. A small flatfish, the speckled sanddab (*Citharichthys stigmaeus*) shakes off its powdery cover and rises just above the sand. Like an undulating cloak, the sanddab flutters away, then hovers and discreetly settles down again. Mottled to match the colors of the sand, it lies flat against the bottom, left-side-up. Its camouflage is nearly perfect as it rests unseen, bulging eyes scanning for worms, shrimps and fishes.

Outstretched like a feather above its sandy burrow, a sea pen filters plankton from the water.

Electric organs on either side of a Pacific electric ray's head (above) generate stunning jolts of electricity that immobilize fishes.

In this series, a rainbow nudibranch approaches a burrowing anemone (top). The hungry nudibranch attacks the anemone tentacles (middle). In quick defense, the anemone contracts into its tube, dragging the predator in with it (bottom).

5

SEASONS IN MONTEREY BAY KELP FORESTS

As the seasons progress, kelp forest residents adjust to the sunny calm of spring or the violent storms of winter and the seasonal changes in their neighborhoods. Each year brings different patterns of weather, disease and competition for food and space in the forest. In mild winters, there may be minor storm damage to the forests. But a succession of severe winters can be so devastating, it may take years for the forests to recover. Some years bring warmer temperatures, or currents poor in nutrients, or epidemics of bacteria or other disease organisms that ravage the kelp forest. But from time to time, when conditions are right, it's like an enchanted forest. Pioneering spores combine with perfect growing conditions to produce seaweed magic, and unfamiliar species join old favorites to flourish in the undersea garden.

SPRING AND SUMMER The coastal morning fog of our spring and summer seasons are a product of upwelling in Monterey Bay, when cold, nutrient-rich water wells up from the deep. Upwelled water fertilizes the seaweeds and phytoplankton. In the cold, rich waters, tiny diatoms quickly multiply. Each species of these miniature plants has its own fantastic shape: cylinders, oblongs, spheres and crescents. By the hundreds and then thousands, the dividing diatoms launch a frenzied population explosion.

In early spring, stumps of raggy red seaweeds on the forest floor sprout new blades and branches that vigorously grow. New kelp fronds unfurl from the holdfast and stretch out and up toward the surface. These new kelp blades are clean—unspoiled by the bites of grazing fishes or settled organisms. The fronds grow gradually up toward the surface of the water to resurrect the forest canopy.

Water streams through the kelp, carrying a miniature zoo. Microscopic swimming stages of the moss animals drift in. One of them is a tiny larval bryozoan, a clamlike swimmer in a bivalve shell. Opening its shell like an umbrella, it parachutes down onto a clean kelp blade. Alert for chemical cues, the bryozoan tests the surface, then cements itself to the blade with a sticky glue. The youngster settles in place and changes to its adult form, a captive within its own shelled rectangular fort. Extending a crown of tentacles above its shell, the bryozoan flicks its tentacles through the water to catch bits of food. Once established on the kelp, the lone settler begins to multiply. Budding off clones in neat rows, a colony fans out to frost the blade with a crust of the tiny animals.

In the calms of autumn, giant kelp canopies spread in graceful patterns over the water's surface (top).

A red seaweed on the forest floor (above) shines with blue iridescence.

Carried into the bay on ocean currents, these schools of ocean sunfish search for jellies (right).

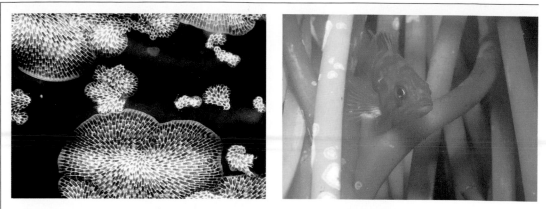

Microscopic crusts of bryozoans expand over the kelp blades (above left).

Juvenile rockfish hide among the kelp stipes (above).

A curious harbor seal observes other visitors to the kelp canopy (left).

In spring and summer, fresh figures navigate the forest channels. Swarms of juvenile rockfishes drift through the canopy, small and defenseless during their first few months. They wander with the plankton, searching out meals of smaller drifting animals. Occasionally, larger juveniles eat smaller ones with cannibal gusto, only to be gobbled up in turn by the adults.

A surfperch swims above the rocky reef.

A loose school of kelp surfperch gathers near the water's surface, while far below, black surfperch (*Embiotoca jacksoni*) and colorful striped surfperch (*Embiotoca lateralis*) swim just above the rocky reefs. These fishes live in many habitats, eating everything from worms and molluscs to parasites on other fishes. The female surfperch that carried litters of young through the winter now deliver two-inch-long newborns that flash over the reef. Some of the newborn males are ready to mate as soon as they're released from their mothers.

The blackeye goby builds a nest and guards its eggs.

In a pocket of sand near the rocky reef, a male blackeye goby (*Coryphopterus nicholsii*) digs out a clean spawning area under a rock. He rises above his nest to display his black pelvic disk, a performance that attracts a female goby to lay her eggs. After fertilizing the eggs, the male goby guards his precious nest until the young hatch.

Harbor seals (*Phoca vitulina*), California sea lions (*Zalophus californianus*) and other marine mammals hang out in the kelp

The cormorant's bright blue throat patch signals that it's mating season (above).

This nursing sea otter pup is nearly as big as its mother (left).

forest. Once in a while, gray whales tour the forests close to shore as they migrate from their breeding sites off Baja California to their northern feeding areas in the Bering Sea.

Mother sea otters coach their offspring on survival skills. One patiently teaches her pup to dive. Together, they plunge through the watery neighborhood to hunt for urchins, abalones and sea stars on the forest floor. Sleek and sure under water, they re-surface for air after two minutes below. At the surface, the otter pup imitates his mother and chews on a sea star; then, finding the star distasteful, tosses it aside and crawls up on his mother's chest to nurse. The protective mother stays alert for sharks or other dangers, but today they're safe in the water.

A skinny sea crow, the Brandt's cormorant (*Phalacrocorax penicillatus*) flies out to the kelp forest to collect its nesting materials. Diving below the water's surface, it paddles down to the bottom, grabs a mouthful of red seaweeds and heads back toward the surface. Carrying the dripping plants to shore, the cormorant arranges the seaweeds into a crude nest. A female joins his nest-making; her bright blue throat pouch signals that she's in breeding season. All around them, male cormorants squabble for territory and couples prepare their nests.

As summer passes into fall, scenes shift, new characters come in and alliances are altered in the kelp forest.

MONTEREY BAY SEASONS

A shroud of morning fog envelops the bay in summer.

The California Current that brings cold water southward along the West Coast dominates the marine seasons in Monterey Bay. These seasons are also complicated by other water masses, so that three distinctly different "oceanic seasons" occur during the year. Although these "seasons" vary from year to year, they generally follow in sequence. The local marine climate, including our summer fog, is directly related to the currents and marine seasons along the coast. As warm, moisture-laden air moves across the Pacific, it encounters the cold surface waters of the California Current and nearshore upwelling, and suddenly cools. Rapid cooling causes the water vapor to precipitate as fog—a common feature along this coast in the summer.

From November through February, the north-flowing Davidson Current brings relatively warm water close to shore. During these months, you'll find a well-mixed sea with the lowest salinities of the year, slow-cooling surface temperatures and little change in temperature from the surface to depths of 160 feet. Gray whales often migrate through the bay at this time of year.

The upwelling season of cold, nutrient-rich water often begins in March with the onset of the prevailing northwesterly winds characteristic of spring and summer. Upwelling often lasts through September. The northwesterly winds drive surface water offshore, allowing cold, deeper water to flow up to surface. Surface temperatures reach their lowest of the year before climbing, in late summer, toward their peak. The ocean waters are stratified: temperatures drop rapidly with depth and vary considerably throughout the bay. Fueled by upwelled nutrients and summer sunlight, blooms of phytoplankton enrich bay waters.

The short oceanic period (generally September and October) brings the highest surface temperatures to the bay. Clear, blue water flows in from offshore as the cold upwelling water begins to sink in the absence of northwesterly winds. Plankton and fishes from well offshore are often seen in the bay during this season.

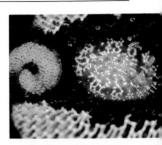

FALL IN THE KELP FOREST Autumn brings the best diving weather to the forest when upwelling ends and warmer water rolls in from offshore. The water is flat calm, clear and blue, and almost warm. The kelp and other seaweeds are still lush from summer, but they grow more slowly as natural fertilizers in the water are used up.

By fall, the kelp canopy looks thick and well-developed, extending out across the water's surface in a tangle. The bryozoans that settled down in spring have fanned out to cover whole kelp blades, but now there's a menace in their neighborhood. A pale, flat sea slug (*Doridella steinbergae*) settles down on the bryozoan colony. Patterned and colored to match the bryozoan, its disguise is nearly perfect. Other predators won't disturb it as it slides over the colony to feed on its immobile host.

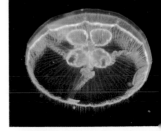

Rivers of transparent ocean water swerve close to shore, carrying mobs of microscopic dinoflagellates and large animal drifters into the kelp forest. The tiny dinoflagellates divide: two, four, eight, sixteen and on and on in geometric series. Weird asymmetric shapes, some release a fiery glow as they twirl through the water. Their flashes of light may warn plankton-eaters that these brilliant performers can be poisonous.

Sculptured bells and balls of gelatin glide through the forest: the jellies and comb jellies. A flat moon jelly (*Aurelia aurita*) dances by, its bell contracting rhythmically. Once a bigger plate, this hungry moon jelly has dwindled to a smaller saucer size. Food arrives as a swarm of buglike copepods bump the jelly and stick to its slime-covered coat. Tiny wiggling hairs on the jelly transport each bit of food to its mouth to fatten up the moon jelly again.

Camouflaged against its bryozoan prey, a flat sea slug lays a curling egg case on a kelp blade (top).

A moon jelly glides through the kelp forest, rhythmically pulsing its bell (middle).

Twinkling with colors, a comb jelly sculls through the water (bottom).

Sculling mouth-first through the water, a comb jelly (*Beroë forskalii*) vibrates with shimmering color. It travels to a rhythmic beat that's driven by eight stripes, each a row of combs made of tiny pulsing hairs. The comb jelly bumps against a smaller jelly and gulps it whole.

The thickened canopy slows the flow of water through the forest. Dozens of adult rockfishes—kelp (*Sebastes atrovirens*), olive (*S. serranoides*) and black rockfishes (*S. melanops*)—pause in suspended animation. They prey on floating larvae and smaller fishes, even gobbling up tiny rockfishes in their haste to feed. Hanging in large loose schools at the edge of the forest, blue rockfish (*S. mystinus*) scoop up jellies and other plankton that drift in from the open sea.

A school of disk-shaped ocean sunfish (*Mola mola*) swim in close to the edge of the kelp forest. Six feet across, these sluggish giants slurp up juicy jellies and other large invertebrates. Some sunfish lie quietly near the surface while señoritas and other cleaner fishes pick off parasites.

Suddenly, a squadron of black birds streaks through the kelp forest. Like synchronized swimmers, these cormorants dive simultaneously to hunt fishes. Legs kicking furiously, the birds descend to attack a school of northern anchovies (*Engraulis mordax*). The orderly school of fish dissolves in panic; streaks of silver scatter, then instinctively regroup. With a successful catch, the cormorants ascend. When they've had their fill of fish, the birds bask on rocks, holding their wings out to dry in the sun. The young cormorants that hatched in spring are already independent. Most have gone their separate ways, some heading far to the north to fish in Puget Sound.

A female sea otter joins a male in his territory in the kelp forest canopy. Her pup weaned, she's ready to mate again. As the new partners swim together and dive for food, he bumps playfully against her. When they mate, the game gets rougher. He may bite her nose or claw her face as they tumble in the water. For a few days they're inseparable, then she breaks the bond and leaves him. She'll bear a new pup late in winter.

When sea otters mate, the male grips or bites the female's nose as she lies on top of him, head bent back.

WINTER IN THE KELP FOREST As fall slips into winter, the kelp forest takes on a tired, senescent look. With shorter days and gray, rainy weather, the seaweeds weaken and decline in the infertile waters. But it's the winter storms that really shake up the forest. Cold winds blow anarchy and chaos into the community. Storms from as far away as the Gulf of Alaska send out huge, destructive swells that reach Monterey Bay even when it isn't stormy here. The raging winds can produce white-capped waves 15 feet tall or more, delivering 40 times the force of hurricane winds on land. The swells disrupt life in the kelp community, shaking animals from their homes and dragging away plants and animals.

The older kelp blades are heavy with bryozoans and hydroids. Some blades wear thick crusts of the white lace, a burden that blocks sunlight and weakens these venerable plants. Sea slugs and kelp crabs graze on the overgrown hydroid gardens, leaving stubby meadows in their wake. In spring, the kelp crabs were vegetarians; now in winter, they eat animals, too. One kelp crab picks at the network of hydroids, then spies larger prey, a leafy sea slug (*Melibe leonina*). As the crab creeps over to harass its new target, the sea slug escapes. It sails into the water, swimming upside down and flexing from side to side.

The elderly kelp blades begin to decay, to the delight of the smallest citizens of the canopy. Snails, amphipods and worms nibble on the weakening blades. Hard at work, a tiny female

amphipod (*Cymadusa uncinata*) rolls the edge of the kelp blade into a tunnel home, all the while carrying a mate on her back. Looking like a narrow pill bug, a brown isopod (*Idotea resecata*) gnaws on the stipe near the float. At last, the blade breaks loose, carrying its microscopic menagerie away to other food webs.

In winter, the red seaweeds show their age and the weight of their summer activities. The narrow blades of *Rhodymenia* are coated with bryozoans. A heavy crop of red sea grapes overloads the branching sea grape stalks. The rough surge will prune many of the red seaweeds from the carpet. A sunflower star rambles through the weeds. It bumps against an abalone and recognizes food as quickly as the abalone recognizes trouble. The big snail twists its shell violently back and forth and "gallops" away before the sunflower star can recover.

The slender clingfish that added mystery to the kelp columns in summer are even harder to find in winter. But surfperches flicker about, courting and mating, while other kelp forest fishes prepare for winter spawn. A female lingcod lays a football-sized clump of eggs in a crevice in the reef; then the male lingcod fertilizes them. He'll stand guard over the nest until all the eggs hatch out.

Female cabezons prepare their massive spawn. Each lays 50,000 eggs in sticky, green masses on the rocks. The smaller male cabezon fertilizes the eggs, then perches nearby. (His guard may be

Storm waves (top) tear through the kelp forest.
A top snail grazes on tattered kelp (middle).
Clinging to a stipe, a transparent Melibe *filters plankton (above).*

redundant; the eggs are toxic, so most animals leave them alone.)

It's also courtship time for kelp greenlings (*Hexagrammos decagrammus*). Orange freckles on one kelp greenling signal that she's a female. Her mate looks distinctly different; he's smaller with bright blue spots on his head for attracting female attention. His sex appeal is evident; the female advances to his chosen rock. In a rocky crevice, she lays a mass of pale blue eggs. Hovering over the eggs, the male kelp greenling releases his cloud of sperm.

Most of the cormorants that fished in the kelp forest during the summer have flown off for the winter. When it's stormy, these birds can't dive for fishes in the kelp forest. Weak with hunger, one lone cormorant huddles on the rocks near shore. The starving bird will be dead before spring. Its body will feed scavenging animals on the beach.

The rough seas of winter have separated many of last year's sea otter mothers from their weaned pups. A male otter patrols its territory, splashing and kicking to advertise his presence to other males in the area. Ignoring his antics, four young females socialize and play in the kelp forest. A pregnant female otter joins the other females. Surrounded by kelp, she'll bear her new pup in the winter waters.

Fewer snails climb up and down the kelp stipes. Every so often, a kelp crab scrambles up and down the columns. Many snails and crabs lose their footing with the sway of huge swells. But the invertebrate commuters have left their marks up and down these vertical turnpikes. The stipes themselves look tired and worn. Some maintain a few beaten and frayed blades where fishes can hide. Most are now bare, knotty cables, stretching bladeless and bedraggled toward the surface.

The shrimps, crabs, gribbles and worms that wiggle and chew on the inside of the holdfast have weakened their home's grip. The younger and healthy holdfasts maintain their firm hold, but the older anchors stagger under winter's heavy swells. In stormy weather, these insecure holdfasts endanger the whole community. Intermittent tugs from surface waves find a weak link in the anchor; two branches in the holdfast stretch thin and finally snap. The holdfast buckles, strains to bolster its position and surrenders. Torn loose from the bottom, the holdfast shakes violently. Many snails and crabs and worms jump ship immediately, throwing their fate to the wind and waves. But locked in the holdfast and stuck on the stipes, hundreds of unwilling prisoners are carried off.

The liberated kelp frond entangles other kelps that are still attached. The snarl of holdfasts, stipes and blades traps other innocent victims. The pull of added weight strains the limits of the healthiest holdfasts. Like a tumble of dominoes, one after another tears free, leaving a swath of devastation in the kelp forest.

Some torn-up kelp plants drift along the floor of the kelp forest. Their floats are punctured, all buoyancy forgotten. High-speed currents sweep this kelp out into the bay and down a submarine canyon. Six hundred feet below the surface, a fragile

A tangle of giant kelp lies in disarray on the shore (top).

Beach hoppers, flies and beetles find a haven in this drying wrack of bull kelp (above).

deep sea urchin (*Allocentrotus fragilis*) captures a meal of kelp from the mass that tumbled downward.

Revealed by winter's incandescent light, fishes swim through the denuded kelp forest.

Buoyed by gas-filled bladders, another huge mass of kelp plants bobs to the surface and drifts toward shore. Thrown ashore, then dragged back down the beach into the water, the plants are thrashed and pummeled by the waves. Sand scrapes and abrades all surfaces. The beaten drift is heaped in masses on the beach, abandoned by the retreating tide. The sea slides back down the beach, still frothy with kelp slime.

A covert beach community delights in this gift of kelp from the sea. Isopods and beach hoppers desert their sandy burrows to feed on the wrack at night. Tiny mites (*Gammaridacarus* sp.) hop on the beach hoppers and amphipods. The mites burrow in the sand by day and forage for food by night. Their favorite food, the wrack of giant kelp, arouses mighty passions. The female mite lays a single pearl of an egg on the moist, decaying wrack, then tucks it in to keep it safe and moist. In sand-colored disguise, rove beetles (*Thinopinus pictus*) hunt the hoppers, while red-legged kelp flies (*Fucellia* sp.) take refuge in the wrack. The flies eat kelp and lay eggs, too. In two short days, the new flies will hatch out.

After the winter storms, the future looks bright again for kelp forest creatures. On the beach, the kelp withers, providing one last beach resort for animals in the sand. Under water, young plants that had been banished to the shadows in fall now bask in the sunlit clearings. Lean and bare, the impoverished kelp forest community awaits its springtime renovation.

6

A DEEPER INTEREST IN KELP FORESTS

This tale of kelp forest communities weaves through the seasons, changing from year to year and from place to place. What is it about the kelp forests that captures our interest?

Perhaps it's the bountiful harvest we reap from these undersea gardens. Kelp forests harbor the young of many valuable plants, invertebrates and fishes. Here, both commercial and sportfishing industries gather a rich harvest of urchins, abalones and lobsters, kelp bass, rockfishes and other nearshore fishes.

We even harvest the kelp itself for its rich lodes of iodine, potassium and other minerals and vitamins. A renewable resource, the kelp keeps on growing when the upper canopy is cut off. Younger fronds sprout from its holdfast, quickly replenishing the kelp canopy.

For hundreds of years, people have collected kelp for food. At one time, it was a source of potash for gunpowder, while today it's mainly harvested for its natural gel called algin. Algin absorbs huge quantities of water, making it ideal as an emulsifying and binding agent in foods like ice creams and puddings, as well as pharmaceuticals, paints and make-up. The foam on your beer may be stabilized by algin. Paper-making and textiles depend on it; it's even in the coatings of welding rods.

But kelp forests are more than a commodity. These majestic underwater paradises draw great numbers of divers who endure bone-chilling cold to drift among the swaying kelp and contemplate the multitude of forest mysteries, the little-known species, secret lifestyles and intriguing relationships of a kelp forest community. From the delicate beauty of a new unfurling frond to the subdued sheen of antique blades, kelp catches the eye and elevates the spirit.

Challenging questions and the beauty of the forests compel students and scientists to explore deeper, to probe the forests and examine their structure and organization. Some target large-scale relationships and ecological questions. What are the seasonal changes in a kelp forest community? How do communities differ at various sites? Monterey Bay Aquarium researchers are comparing succession and the development of the community of plants and animals in the Kelp Forest exhibit with natural sites in Monterey Bay. From airplanes, we're photographing California coastal areas to gather information about seasonal and long-term changes in the forest canopies. Electronic instruments track wave heights and sea temperatures in the bay. Scientists monitor kelp growth in the exhibits and in the wild, and compare kelp survival in areas

Floats resembling perfect spheres (above) provide a contrast to the more commonly seen oval floats in Monterey Bay. No one knows if genes alone can fix the form or if environment sets the shape.

In a southern California kelp forest, this solitary urchin awaits a meal of drift kelp (right).

THE KELP FOREST UP CLOSE

The Kelp Forest exhibit at the Monterey Bay Aquarium represents the first successful attempt to grow and maintain a giant kelp community in any aquarium. The exhibit tank, two stories tall and open to the air and ambient sunlight, is 28 feet tall and 66 feet long and holds 330,000 gallons of sea water. The clear acrylic panels—15 feet long and eight feet wide—are about 7-½ inches thick.

Sea water pumped in from the bay ensures a constant supply of nutrients for the giant kelp and other seaweeds and a surge machine at the top of the tank keeps the water moving. By day, the incoming sea water is filtered so the water is clear in the exhibit. By night, unfiltered sea water is pumped into the exhibit, rich with invertebrate larvae and algal spores that may settle and grow in the exhibit or provide food for filter-feeding animals in the exhibit.

The living giant kelp and other seaweeds provide habitat, shelter and food for numerous fishes (like sheephead, kelp bass, rockfishes, surfperches, señoritas, leopard sharks, horn sharks and greenlings) and invertebrates (like sea stars, anemones, crabs and abalones). Researchers at the aquarium study succession and seasonal changes in the community. They analyze chemical changes in the seawater and measure growth of the kelps at different times of the year. As additional plants and animals settle and grow in the exhibit, our scientists monitor these developments.

An experiment from its beginning, the Kelp Forest exhibit has become a living laboratory where we can observe and study life in a swaying, growing kelp forest.

Visitors watch a diver examining kelp plants inside the Kelp Forest exhibit.

devoid of sea urchins with areas that have been deforested by grazing sea urchins.

Some research focuses on individual organisms or on interactions between small groups, like competition between closely related species of surfperch; reproduction and dispersal of a particular kind of barnacle; or the longevity of individual kelp plants. Each frond of giant kelp may survive for six months, and individual blades may last only four months, but the plant itself is perennial, living for a number of years. Over time, the fronds may mature in place, slough off tissue, or break off and drift away. Kelp litter doesn't accumulate in the forest, so where does the bulk of kelp go? Do the forests export material to support communities elsewhere, such as on the beach or deep in submarine canyons?

Scientists use radio transmitters to track kelp drifting on the surface of Monterey Bay. Deep sea instruments, like remotely operated vehicles (ROV) and the manned submersible, *Alvin*, help researchers photograph and study the abundance of drift kelp in the deep submarine canyons just off the coast of Monterey. Seaweed traps set 1,000 feet below the surface of the bay retain drift kelp for months so we can examine the microorganisms (like bacteria and fungi) that break kelp down and learn how long it takes kelp to decompose in deep water.

Given the variety of shapes and sizes in the kelp forest community, scientists want to know about the forces of flowing water on plant and animal structures. How do eggs and sperm come together in a free-flowing environment? How does a spore or larva

Large ships like this kelp harvester prune the kelp canopy, leaving enough submerged kelp plants behind to grow a new canopy (top left).

A researcher climbs out of the submersible Alvin after exploring the Monterey submarine canyon (above left).

Two scuba divers prepare for their descent into the kelp forest (above).

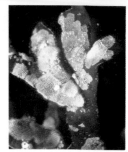

This diver surveys the dramatic panorama as he descends into the kelp forest (left).

A few animals can change the face and future of an entire community. Here, a snail devours a kelp blade (top).

A red seaweed cloaked in bryozoans (above) inspires research questions: how do these shrouds affect plant health and longevity?

settle in flowing or turbulent water? The answers to these questions demand both a physical and chemical approach. The speed and mixing of the water determines whether a particle gets close, but chemicals play a role in the attraction of kelp sperm to eggs and in the settling behavior of the red abalone on the pink coralline crusts it calls home.

Some researchers spotlight fine-scale phenomena, looking inside kelp forest organisms at their cells, organelles and molecules. We use biochemistry, physiology and molecular biology to study how organisms are related and how they're different. Kelp specialists may compare the physiology of different stages of kelp: the large sporophytes, diminutive gametophytes, and minute spores and sporelings. How does photosynthesis differ in the different stages? Is there competition between the tiny gametophyte stages of the kelp? How does shading from the giant kelp canopy affect the tiny plants on the forest floor? Chemical analyses of seaweeds may relate plant growth and reproduction to seasonal changes in sunlight or to the concentrations of nutrients in the sea water.

Research in the kelp forest leaps from every direction. On the grand scale, we're celebrating the diversity of life in the kelp forest and we marvel at how such a variety of organisms live and work together. On the fine scale, we're examining minute parts of one particular creature. On every level, the kelp forest presents multitudes of unanswered questions. The work of understanding kelp forests has just begun. The challenge is still before us.

A Sampling of Monterey Bay Kelp Forest Species

Seaweeds

Acid seaweed *Desmarestia ligulata*
Brown seaweed *Cystoseira osmundacea*
Bull kelp *Nereocystis luetkeana*
Coralline, jointed *Bossiella* spp.
Coralline, jointed *Calliarthron* spp.
Giant kelp *Macrocystis pyrifera*
Red seaweed *Fauchea* spp.
Red seaweed *Rhodymenia* spp.
Sea grapes, red *Botryocladia pseudodichotoma*
Turkish towel seaweed *Gigartina* spp.
Understory kelp *Eisenia arborea*
Understory kelp *Pterygophora californica*

Invertebrates

Abalone *Haliotis* spp.
Amphipod *Cymadusa uncinata*
Anemone, aggregating *Anthopleura elegantissima*
  strawberry *Corynactis californica*
  tube-dwelling *Pachycerianthus fimbriatus*
Bat star *Patiria miniata*
Brittle star *Ophiothrix spiculata*
Bryozoan *Membranipora tuberculata*
Comb jelly *Beroë forskalii*
Crab, decorator *Loxorhynchus crispatus*
  hermit *Pagurus samuelis*
  rock *Cancer antennarius*
  kelp *Pugettia* spp.
Gumboot chiton *Cryptochiton stelleri*
Hydroid *Eucopella* sp.
Isopod *Idotea* spp.
Melibe *Melibe leonina*
Moon jelly *Aurelia aurita*
Octopus, red *Octopus rubescens*
Opossum shrimp *Acanthomysis* sp.
Sea slug *Doridella* sp.
Sea star *Pisaster giganteus*
Sea urchin, purple *Strongylocentrotus purpuratus*
  red *Strongylocentrotus franciscanus*
Sponge, blue *Hymenamphiastra cyanocrypta*
Sunflower star *Pycnopodia helianthoides*
Top shell *Calliostoma* spp.
Tube worm *Diopatra splendidissima*
Turban snail *Tegula* spp.

Fishes

Blacksmith *Chromis punctipinnus*
Cabezon *Scorpaenichthys marmoratus*
Garibaldi *Hypsypops rubicundus*
Giant kelpfish *Heterostichus rostratus*
Halfmoon *Medialuna californiensis*
Kelp bass *Paralabrax clathratus*
Kelp greenling *Hexagrammos decagrammus*
Kelp gunnel *Ulvicola sanctaerosae*
Kelp surfperch *Brachyistius frenatus*
Lingcod *Ophiodon elongatus*
Opaleye *Girella nigricans*
Painted greenling *Oxylebius pictus*
Rockfish, black *Sebastes melanops*
  black-and-yellow *Sebastes chrysomelas*
  blue *Sebastes mystinus*
  gopher *Sebastes carnatus*
  kelp *Sebastes atrovirens*
  olive *Sebastes serranoides*
Señorita *Oxyjulius californica*
Sheephead *Semicossyphus pulcher*
Torpedo ray *Torpedo californica*
Wolf-eel *Anarrhichthys ocellatus*

Birds

Brandt's cormorant *Phalacrocorax penicillatus*
Eared grebe *Podiceps nigricollis*
Elegant tern *Thalasseus elegans*
Horned grebe *Podiceps auritus*
Surf scoter *Melanitta perspicillata*
Western gull *Larus occidentalis*

Marine Mammals

Gray whale *Eschrichtius robustus*
Harbor seal *Phoca vitulina*
Killer whale *Orcinus orca*
Sea otter *Enhydra lutris*
Sea lion *Zalophus californianus*

INDEX